SPOTLIGHT ON GLOBAL ISSUES

CLEAN WATER FOR ALL

Danielle Haynes

ROSEN PUBLISHING
NEW YORK

Published in 2022 by The Rosen Publishing Group, Inc.
29 East 21st Street, New York, NY 10010

Copyright © 2022 by The Rosen Publishing Group, Inc.

All rights reserved. No part of this book may be reproduced in any form without permission in writing from the publisher, except by a reviewer.

First Edition

Editor: Theresa Emminizer
Book Design: Michael Flynn

Photo Credits: Cover Riccardo Mayer/Shutterstock.com; (series globe background) photastic/Shutterstock.com; p. 5 Lukasz Pawel Szczepanski/Shutterstock.com; p. 6 Marco Ramerini/Shutterstock.com; p. 7 FOTOGRIN/Shutterstock.com; p. 8 Fouad A. Saad/Shutterstock.com; p. 9 Chris Overby/Shutterstock.com; p. 10 Smith Collection/Gado/Archive Photos/Getty Images; p. 11 Fratelli Alinari IDEA S.p.A./Corbis Historical/Getty Images; p. 12 Stephen J. Boitano/Getty Images; p. 13 Andrea Izzotti/Shutterstock.com; p. 15 Aaron P. Bernstein/Getty Images; p. 16 Riccardo Mayer/Shutterstock.com; p. 17 Martchan/Shutterstock.com; p. 18 paula french/Shutterstock.com; p. 19 Hindustan Times/Getty Images; p. 21 Rich Carey/Shutterstock.com; p. 22 Pcess609/Shutterstock.com; p. 23 MediaNews Group/Getty Images; p. 25 arindambanerjee/Shutterstock.com; p. 27 Richard Drew/AP Images; p. 29 Pantiwa Lakum/Shutterstock.com.

Cataloging-in-Publication Data

Names: Haynes, Danielle.
Title: Clean water for all / Danielle Haynes.
Description: New York : Rosen Publishing, 2022. | Series: Spotlight on global issues | Includes glossary and index.
Identifiers: ISBN 9781725323049 (pbk.) | ISBN 9781725323070 (library bound) | ISBN 9781725323056 (6 pack)
Subjects: LCSH: Water-supply--Juvenile literature. | Water quality--Juvenile literature. | Water--Pollution--Juvenile literature. | Water conservation--Juvenile literature.
Classification: LCC TD348.H39 2022 | DDC 363.6'1--dc23

Manufactured in the United States of America

Some of the images in this book illustrate individuals who are models. The depictions do not imply actual situations or events.

CPSIA Compliance Information: Batch #CSR22. For further information contact Rosen Publishing, New York, New York at 1-800-237-9932.

CONTENTS

THE BLUE PLANET 4
KEEPING THINGS MOVING 8
CLEANING THINGS UP. 10
DISASTER STRIKES. 14
FINDING A SOLUTION 18
PLASTIC IN THE WATER 20
TRIBAL WATER RIGHTS 24
WATER WARRIORS. 26
WATER STEWARDSHIP 30
GLOSSARY 31
INDEX 32
PRIMARY SOURCE LIST 32
WEBSITES. 32

CHAPTER ONE

THE BLUE PLANET

Most people know that Earth's surface is largely covered by water. In fact, 71 percent of Earth's surface is covered by H_2O. If you were to gather all the planet's water into a sphere, it would be 860 miles (1,384 km) in diameter, about the distance from Salt Lake City, Utah, to Topeka, Kansas. It's no wonder Earth is known as the Blue Planet!

Knowing this, it seems there should be more than enough water for drinking, bathing, and irrigating crops. But in reality, finding enough clean, drinkable water for all 7.6 billion people on Earth can be a challenge.

Because water is found in so many connected places on Earth—in streams, rivers, creeks, lakes, and oceans—it's very susceptible to contamination. This means that if there's a chemical spill near a river, the toxins could travel miles downriver, trickle through connecting streams, or even flow into the sea.

Dirty water isn't the only problem. Billions of people across the globe don't have access to any water at all. This is because they live either in places that have droughts or places where the local government doesn't make enough water available to them.

Every person and animal on Earth needs clean water to live. You may not think much about where the water from your faucet comes from or whether it's clean enough to drink, but it's an issue that affects every single one of us. Luckily, there are many ways people can come together to reach the goal of providing clean water for all.

There are about 326 million cubic miles (1.4 trillion cubic km) of water on Earth, enough to make the planet appear mostly blue from outer space.

Humans have long recognized the importance of sanitary, or clean, water. When ancient people first shifted from nomadic, or traveling, lifestyles into permanent settlements, they began building wells for fresh water. They also built toilets to keep the wells from being contaminated by waste.

They knew there are an endless number of ways water can become unclean. Even though there are trillions of gallons of water on Earth, that water must be free of toxins and disease to be safe for humans to drink and for animals and plants to live well.

Ancient Greeks and Romans knew to drink tasteless, odorless, and colorless water, and they knew to avoid water from mountains where there was mining. Some historians believe ancient peoples such as the Sumerians and Egyptians even drank beer to avoid contaminated water from rivers and canals.

Chemicals, waste, or microorganisms, all of which can be dangerous or even deadly for humans, animals, and plants alike, can pollute water. Water is especially susceptible to contaminants because it's what's known as a "universal solvent," meaning it can dissolve, or break down, most substances on Earth.

Toxins can enter Earth's waterways as runoff from factories and leaks or spills from oil pipelines or tankers. Flooding can wash pesticides and fertilizers from farms into nearby rivers.

Chemicals aren't the only problem. Bacteria from human and animal feces, or waste, can spread diseases such as typhoid fever, cholera, and dysentery into waterways or crops in the fields. This is how the bacterial infections E. coli and salmonella can contaminate vegetables sold at the store.

Some of the oldest-known water wells have been found on the island of Cyprus. They date back 9,000 to 10,500 years to during the Stone Age.

CHAPTER TWO

KEEPING THINGS MOVING

Ancient civilizations were built around access to clean, drinkable water. Even though there are trillions of gallons of water on Earth, the vast majority—96.5 percent—is salt water and undrinkable. Of the remaining 3.5 percent of water that's fresh, about 69 percent is inaccessible, frozen in the polar ice caps. So finding or making fresh water is critical.

Early efforts to obtain safe drinking water often consisted of ways to bring clean water to where people needed it. Before the third century BC, Egyptians may have invented a device called a water screw—later called the Archimedes' screw after the ancient Greek engineer who brought the technology to his home country. The water screw is a circular pipe with an enclosed helix shape. When inclined and turned, it can move water from one elevation to another.

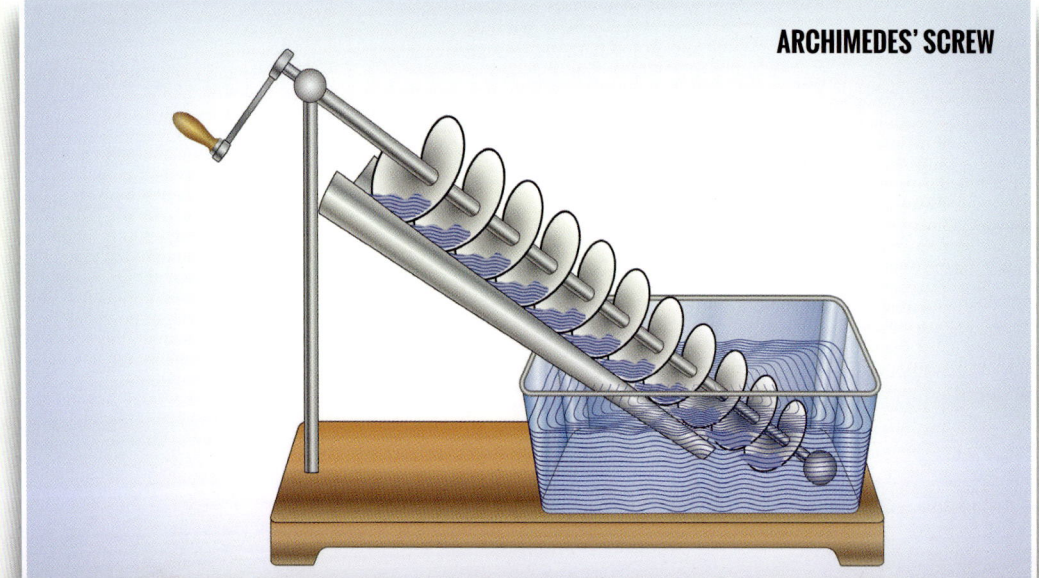

ARCHIMEDES' SCREW

Ancient Egyptians used it to take water from the Nile River to irrigate their crops. Today, sewage treatment plants use Archimedes' screws to move wastewater because they don't tend to become clogged by solid waste.

Ancient Romans developed aqueducts to transport water dozens of miles. An aqueduct is a system of underground pipes, ditches, and canals and aboveground arched bridges that uses gravity to bring water from a higher elevation to an area of lower elevation. From the fourth century BC to about the third century AD, the Romans built 11 aqueducts, one transporting water as far as 57 miles (91.7 km).

These systems brought fresh drinking water to cities where it was hard to find. They were also used to flush wastewater away.

Many of the ancient Roman aqueducts fell into disuse over the years, but you can still see the tall, aboveground arched bridges of these structures in modern-day France, Spain, and Italy.

CHAPTER THREE

CLEANING THINGS UP

Even though ancient people developed ways to gain access to clear and odorless water, they weren't always aware it might not be free of contamination.

While we may think of water treatment plants as an entirely modern technology, humans found ways to clean their water as far back as 4000 BC. Archaeologists have found ancient Sanskrit and Greek writings mentioning methods of purifying, or cleaning, water such as filtering it through charcoal, boiling it, exposing it to sunlight, and straining it. By about 1500 BC, ancient Egyptians learned to use the chemical alum to cause particles to settle out of water.

WATER FILTRATION SYSTEM
STILLWATER, MINNESOTA, 1912

Many years later, in the 19th century AD, scientists Louis Pasteur, Joseph Lister, and Robert Koch discovered that certain diseases spread to humans through microorganisms in the air and water. They called this scientific breakthrough "germ theory."

It was around the same time that an increase in toxins produced by the Industrial Revolution started polluting our waters even more. Combating these potentially deadly contaminants inspired scientists and engineers to develop more advanced forms of filtration and treatment.

In 1804, Scottish engineer John Gibb invented a sand filter process that was so successful that it became law to treat all water from a certain section of the River Thames in England with it.

Today, most U.S. cities use the chemical chlorine to disinfect and kill any germs we can't see in our drinking water. Though chlorine can be harmful, it's used in small enough amounts to kill bacteria but not hurt people.

The first U.S. city to use chlorination as a water treatment method was Jersey City, New Jersey, in 1908.

Today, scientists are constantly working to invent new and more efficient ways to clean water. As climate change increases the severity of droughts across the globe, more people must rely on desalination—the process of removing salt from ocean water to make it safe to drink.

But removing salt from water on a large scale takes a lot of energy and can be costly. It can cost up to $2 to desalinate 264 gallons (999.4 L) of ocean water, which is about how much water a person in the United States uses in two days. The same amount of water from a river or aquifer costs 10 cents to 20 cents.

As new scientific discoveries are made—sometimes in completely different fields—researchers are able to apply the new information to clean water technology. For example, in 2003, scientist Peter Agre won the Nobel Prize in Chemistry for discovering aquaporins. These are proteins in cell membranes that filter water in our bodies. A few years later, researchers at the University of Illinois developed a synthetic, or man-made, membrane with these aquaporins with the idea that it could be used to desalinate and clean water on a larger scale.

PETER AGRE

Other methods of desalination include reverse osmosis, which forces water through filters to remove the salt, and distillation, in which salt water is boiled and the clean water vapor is collected.

But desalination leaves behind concentrated waste, which treatment plants must find a safe way to store. Returning it to the ocean could harm the ecosystem.

As of 2019, there are more than 20,000 desalination plants worldwide. More than 300 million people rely on water from these plants for some or all of their daily needs.

CHAPTER FOUR
DISASTER STRIKES

Despite all the technologies developed to ensure clean drinking water for people, sometimes the unexpected happens and people and the environment can be exposed to unsafe water.

One of the biggest environmental disasters to affect the planet was a massive oil spill in the middle of the Gulf of Mexico in 2010. An explosion on the Deepwater Horizon oil rig released between 134 million and 206 million gallons (507.3 million and 779.8 million L) of oil into the ocean over the course of five months. Eleven workers on the rig died in the initial explosion.

Oil washed up on beaches in Louisiana, Mississippi, Alabama, and Florida and killed a variety of animals, including more than 1 million birds and 65,000 turtles. Because the oil contaminated so much sea life, fishing in these waters was banned to prevent humans from consuming the toxins.

Sometimes the effects of a water disaster can't be seen right away. In 2014, the city of Flint, Michigan, changed the source of its water supply from the Detroit water system to the Flint River. City officials said the water was safe, but as time passed, residents began noticing the water smelled and tasted bad, and people started reporting health problems. More than a year later, doctors began to find high levels of lead in the blood of children living in Flint.

When the city switched its water source, the new water was more corrosive than the old water. The new water caused lead in the city's pipes to leach into the water supply, sickening thousands of people.

Amariyanna Copeny is a 12-year-old activist who is speaking out about the water crisis in Flint, Michigan. She has become known as "Little Miss Flint."

Some people don't have access to the technology to keep water clean. They live in places where the government hasn't invested in sanitation infrastructure, or equipment and structures. This means there may not be pipes in place for bringing in fresh water and moving out wastewater. There may not be water purification plants, so toxins or diseases may contaminate any water that's available. This results in water scarcity, or shortage.

Water scarcity usually happens in places under extreme drought. But even places where plenty of water is available can have problems finding access to clean water if the proper systems aren't in place. Africa has some of the worst water stress in the entire world because of drought, supply problems, contamination, and even violence.

For example, the war in the Darfur region of Sudan during the last few decades can in large part be blamed on a conflict between Arab livestock herders and Black farmers over access to clean water.

Water scarcity can place a heavier burden on women and girls in these communities. In some of these places, women are responsible for finding and obtaining the water their families need for drinking, washing, and cleaning. It could mean walking miles each day just to collect the day's water or finding a safe place to go to the bathroom.

Women and girls collectively spend about 200 million hours each day to find water. They spend another 255 million hours finding a safe place to go to the bathroom. This means many young girls don't have time to go to school.

One in three people in Africa are affected by water scarcity.

CHAPTER FIVE

FINDING A SOLUTION

Water scarcity means it can take a lot of effort to find water, but this issue can also be deadly. Scarcity leads to the spread of deadly diseases, suffering for animal populations, and even famine because of lack of irrigation for crops.

When people have to resort to drinking unclean water—or eating food irrigated by contaminated water—they become exposed to waterborne diseases such as typhoid, cholera, and dysentery. These diseases can cause diarrhea, which causes dehydration. Diarrhea is responsible for 4 percent of deaths globally. In India, where 99 million people lack access to safe water, more than 500 children under the age of 5 die each day from diarrhea.

When people spend so much time sick or traveling to find water, their local economy suffers because they could otherwise be spending their time working or going to school.

Meanwhile, in places of severe drought, the ecosystem suffers too. Wetlands and lakes disappear, eliminating places where animals can drink and give birth to young. In fall 2019, at least 200 elephants died of dehydration in Zimbabwe because the country missed its usual October rains.

Luckily, organizations exist worldwide to find solutions to these problems. The World Wildlife Fund works to educate people and protect wetlands. Charities such as Water.org offer loans to people to help pay for water access. The World Health Organization, which offers education on public health and water quality, says that in 2015, 91 percent of the world's population had access to an improved drinking water source, up from 76 percent in 1990.

Even in places where there's often plenty of rain, such as India, it can be hard to find enough drinkable water because of a lack of supply and sewage infrastructure.

CHAPTER SIX
PLASTIC IN THE WATER

Plastic is a relatively new problem in the age-old battle to find water and keep it clean. Invented around 1900, plastic may seem like a modern solution to many of the world's problems. For example, charities often deliver clean water to water-stressed communities in plastic bottles.

But there's a problem with plastic that its inventors didn't consider—it doesn't biodegrade, or break down, like other waste. The plastic bottle you drank water from could sit in a landfill forever. Plastic can eventually break down through exposure to sunlight (photodegradation) or the weathering motion of water. But even those processes leave behind tiny particles called microplastics.

Ten percent of the waste generated by humans is plastic. But where does it go when we're done with it? Large and small pieces of plastic can both have a harmful effect on the water we rely on. The chemicals from bigger pieces of plastic buried in landfills can leach into the land around them and contaminate groundwater. Those chemicals can make their way into the water we use for drinking and irrigating crops.

Tiny pieces of plastic, some of which we can't even see, can be even more dangerous. Microplastics are pieces of plastic that are 0.2 inch (5 mm) in length or smaller. Some manufacturers make these small bits to add to beauty products and cleansers as an exfoliant. Other microplastics are formed when bigger pieces of plastic degrade over time. They're so small that fish and birds will eat them, passing microplastics into the food chain.

It's pretty obvious that plastic shouldn't be eaten. However, microplastics make their way into water and into the food chain.

In recent years, scientists have discovered that microplastics can be found nearly everywhere—in water, the stomachs of animals, the air we breathe, and the food we eat. Scientists estimate humans ingest, or consume, about 0.18 ounces (5 g) of plastic each week. That's the equivalent of a credit card!

Because the study of how microplastics affect our health is relatively new, doctors aren't sure how harmful they are to humans. But when birds and fish started to be found dead with their stomachs full of the stuff, researchers knew it was time to find a solution to the problem.

One big problem is the Great Pacific Garbage Patch, the largest of several clusters of plastic floating in the world's oceans. There are between 1.15 million to 2.41 million tons (1 million to 2.2 million mt) of plastic in this cluster, which is about twice the size of Texas. It's situated about halfway between Hawaii and California.

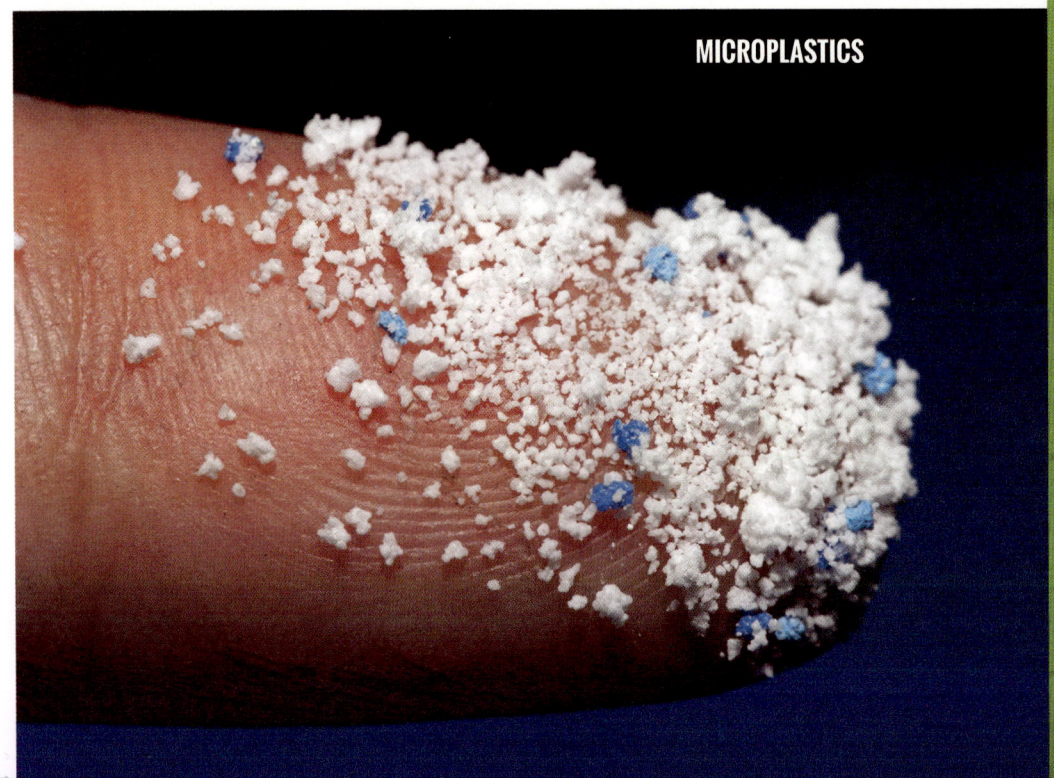

MICROPLASTICS

In October 2019, the Ocean Cleanup's barrier successfully collected plastic from the ocean for the first time.

There are a number of projects worldwide in which people are attempting to clean the ocean of plastic, including the Ocean Cleanup, founded by Dutch inventor Boyan Slat. Trying to collect plastic waste from waterways with boats and nets is time-consuming and costly, so the organization is working to develop more efficient technology. One method involves floating barriers that collect and remove plastics. Another device, called the Interceptor, can be placed in rivers to filter out plastic before it reaches the ocean.

The Ocean Cleanup estimates that once its team of cleanup systems are up and running, it can remove 50 percent of the Great Pacific Garbage Patch every five years.

CHAPTER SEVEN
TRIBAL WATER RIGHTS

In some places, having access to water is a political issue. That's especially the case for Native Americans.

In 1908, the U.S. Supreme Court ruled that Native Americans who were resettled onto reservations by the U.S. government during the 19th century have senior rights to water sources on reservation lands. This means that, for example, non-Native American settlers can't come in and dam up a river, preventing it from running through a reservation.

Many Native American tribes view those water rights as a confirmation of their self-government of their land and a way to make sure they have the water they need to drink and grow crops. Unfortunately, the 1908 case wasn't entirely clear and it hasn't always been enforced, or carried through. This has led to many legal battles today.

Similarly, the 1868 Fort Laramie Treaty established the rights of the Sioux Nation over reservations in parts of South Dakota, North Dakota, Wyoming, and Nebraska. Native Americans from one of those reservations, Standing Rock, held protests and sued the government over the construction of an oil pipeline in 2016 and 2017. They argued that the Dakota Access Pipeline posed a risk to the reservation's natural and cultural resources, including water. If the pipeline leaked, oil could contaminate their water supply, the Missouri River.

Thousands of Native American water protectors and their supporters—including celebrities—set up camps in North Dakota to try to block the pipeline's construction. Their lawsuit failed and the pipeline became operational in June 2017.

Members of the Sioux Nation and thousands of their supporters, including groups of veterans and celebrities, protested the Dakota Access Pipeline.

CHAPTER EIGHT
WATER WARRIORS

Luckily, there are many people in the world willing to advocate for clean water. Pioneers such as Dutch inventor Boyan Slat and American researcher Peter Agre choose to use science and technology to solve the problem of contaminated water. Others, such as actors Matt Damon (cofounder of Water.org) and Shailene Woodley (Dakota Access Pipeline protester), use their celebrity status to draw attention to the issue.

Some, like Canadian teen Autumn Peltier, become a voice for their community at a very young age. Autumn, a member of the Wikwemikong First Nation, has advocated for clean water for indigenous communities since she was 8 years old. She has been called a "water warrior." In fact, the Anishinabek Nation, a political organization that includes the Wikwemikong, named Autumn its chief water commissioner when she was just 14.

She has spoken about clean water on some of the biggest stages in the world. In 2016, the then-12-year-old publicly told Canadian Prime Minister Justin Trudeau she was disappointed with policies allowing oil pipelines to threaten waterways. On World Water Day in 2018, she spoke at the United Nations General Assembly in New York City. She urged world leaders to "warrior up" to give water the same protections as humans.

"Many people don't think water is alive or has a spirit," she said. "(But) my people believe this to be true. No one should have to worry if the water is clean or if they will run out of water."

In 2019, Autumn Peltier was nominated for the third time for the International Children's Peace Prize given by the David Suzuki Foundation.

People living in certain parts of the world may not have to worry about access to clean water in their daily lives. They can turn on a tap in their kitchen or bathroom and have an instant flow of safe, affordable water. But people from every corner of the globe should practice water conservation.

Water is a limited resource, and even though it seems like there may be plenty in one community one year, there could be a drought the next year. There are some simple ways everyone can work to limit their use of water.

- Turning off the tap while brushing your teeth could save 8 gallons (30.3 L) each day.
- Taking a shower uses 10 to 25 gallons (37.8 to 94.6 L) of water compared to up to 70 gallons (264.9 L) for a bath.
- Keep showers under five minutes long.
- Use low-flush toilets and low-flow shower heads.
- Run your dishwasher and washing machines only when you have a full load of dishes or clothes to clean.
- Avoid watering your lawn in the hottest part of the day to keep the water from evaporating right away.
- Water your lawn only every three to five days, and use a timer on your sprinklers to avoid overwatering.
- Fix leaks on any taps or other water connections inside and outside your home.
- Use a bucket of water and sponges to clean your cars and bicycles, not a running hose.

Turn off the tap while you brush your teeth. This can save gallons of water each day.

CHAPTER NINE
WATER STEWARDSHIP

Water is one of the most important natural resources. Clean water is necessary for all plants, animals, and people on Earth. Unfortunately, water is susceptible to many forms of contamination, including bacteria, diseases, toxins, and even plastic, any of which could be deadly.

That's why it's important for people to be good water stewards. Water stewardship means taking care of the earth and being mindful about how decisions you make today could affect water quality across the globe.

You can do this in little ways at home. Practice the good conservation skills you learned in this book, reducing the amount of water you and your family use each day. How about cutting down on your use of plastic, which can end up in landfills and the ocean? Try using a reusable water bottle instead of disposable plastic, and stop using plastic straws, grocery bags, and forks and knives.

You could even think a little bigger. Try joining or contributing to an organization that advocates for clean water and assists countries with limited access. Water.org, the charity cofounded by actor Matt Damon, uses fundraising dollars to bring water and sanitation to communities lacking them. You could even start your own water conservation club at school with the goal of educating your peers and encouraging good stewardship on campus. If you want your voice to be heard by an even larger audience, talk to your parents about joining an environmental protest or march.

GLOSSARY

advocate (AD-vuh-kayt) To argue for or support a cause or policy.

aquifer (AH-kwuh-fur) A layer of rock or sand that absorbs and holds water.

commissioner (kuh-MIH-shuh-nuhr) Someone who is put in charge of something.

contamination (kun-ta-muh-NAY-shuhn) The process of making something dangerous, dirty, or impure by adding something harmful to it.

corrosive (kuh-ROH-siv) Capable of wearing away or destroying over time.

diameter (dy-AA-muh-tuhr) The measurement across the center of a round object.

drought (DROWT) A period of time during which there is very little or no rain.

efficient (ih-FIH-shuhnt) Done in the quickest, best way possible.

exfoliant (ehx-FOH-lee-uhnt) Something rough used to smooth a surface.

famine (FA-muhn) A shortage of food that causes people to go hungry.

helix (HEE-liks) Something that has a spiral shape.

indigenous (in-DIH-juhh-nuhs) Having started in and coming naturally from a certain area.

Industrial Revolution (in-DUH-stree-uhl reh-vuh-LOO-shuhn) An era of social and economic change marked by advances in technology and science.

leach (LEECH) To draw out or remove.

microorganism (my-kroh-OHR-guh-nih-zuhm) A very tiny living thing.

pesticide (PEH-stuh-syd) A poison used to kill pests.

reservation (reh-zer-VAY-shun) An area of land set aside by the government for Native Americans to live on.

sewage (SOO-ihj) Waste.

susceptible (suh-SEP-tuh-buhl) Vulnerable to an action or process.

technology (tek-NAH-luh-jee) A method that uses science to solve problems and the tools used to solve those problems.

INDEX

A
Africa, 16, 17
Agre, Peter, 12, 26

C
Copeny, Amariyanna, 15

D
Dakota Access Pipeline, 24, 25, 26
Darfur region, 16
Deepwater Horizon, 14
desalination, 12, 13
drought, 4, 12, 16, 28

F
Flint, 14, 15

G
Gibb, John, 11

Great Pacific Garbage Patch, 22, 23

I
India, 18, 19
Industrial Revolution, 11

K
Koch, Robert, 11

L
lead, 14
Lister, Joseph, 11

M
microplastics, 20, 21, 22

O
Ocean Cleanup, 23
oil, 7, 14, 24

P
Pasteur, Louis, 11
Peltier, Autumn, 26, 27
plastic, 20, 21, 22, 23, 30

S
Sioux Nation, 24, 25
Slat, Boyan, 23, 26
Standing Rock, 24
Sudan, 16
Supreme Court, U.S., 24

W
Water.org, 19, 26, 30
Wikwemikong First Nation, 26
World Health Organization (WHO), 19

PRIMARY SOURCE LIST

Page 11
Harbor in New Jersey. Photograph. January 1, 1890. Corbis via Getty Images.

Page 25
Activists participate in Dakota Access Pipeline protest. Photograph. Jim Watson. December 4, 2016. AFP via Getty Images.

Page 27
Chief Water Commissioner Autumn Peltier addresses the Global Landscapes Forum. Photograph. AP Photo/Richard Drew. Saturday, Sept. 28, 2019. AP Images.

WEBSITES

Due to the changing nature of Internet links, Rosen Publishing has developed an online list of websites related to the subject of this book. This site is updated regularly. Please use this link to access the list: www.powerkidslinks.com/SOGI/cleanwater